L
I
F
E
V
I
E
W
S

Published by Creative Education
123 South Broad Street, Mankato, Minnesota 56001
Creative Education is an imprint of The Creative Company

Art direction by Rita Marshall; Production design by The Design Lab/Kathy Petelinsek
Photographs by Tom Stack & Associates (Erwin & Peggy Bauer, W. Perry Conway, Jeff Foott,
Thomas Kitchin, Joe McDonald, Mark Newman, John Shaw, Spencer Swanger)

Library of Congress Cataloging-in-Publication Data

George, Michael.
Tundra : the barren wilderness / by Michael George.
p. cm. — (LifeViews)
ISBN 1-58341-023-6
1. Tundras—Arctic regions—Juvenile literature. [1. Tundras.
2. Arctic regions. 3. Tundra ecology. 4. Ecology.] I. Title. II. Series.
GB578.99 .G46   2001
577.5'86—dc21   00-045168

First Edition

2   4   6   8   9   7   5   3   1

THE BARREN WILDERNESS
TUNDRA

MICHAEL GEORGE

THE ARCTIC tundra is a world unlike any other. Flat plains stretch as far as the eye can see. Winters are long, dark, and bitterly cold. Only a few hardy animals struggle to survive. But for a few short months every year, the tundra comes to life with unparalleled brilliance. Summer's mild temperatures and continuous **sunshine** support a vivid array of wildlife that flourishes briefly before succumbing to winter once again.

The Arctic tundra covers the northernmost land in Europe, Asia, North America, Greenland, and Iceland. It begins where the great northern **forests** end. Although maps often show a distinct tree line, the forests do not end

The hardy ptarmigan lives on the tundra year-round.

abruptly; the trees gradually become smaller and farther apart. The tundra's marshy plains extend northward to the shores of the **Arctic Ocean**, the frozen sea that covers the top of the world.

The tundra's unusual climate is a result of the earth's movement around the sun. Once each year, the earth makes a complete **orbit**, or circle, around the sun. As the planet makes this annual journey, it is always tilted. For half of the earth's orbit, from March to September, the northern part of the planet is tilted toward the sun. During this period, the tundra is warmed by days of nearly constant sunshine. For the other half of the orbit, from September to March, the northern part of the earth is tilted away from the sun. During this time, **winter** takes hold of the land.

By October, the temperature is well below freezing and the ground is covered with snow. Each day, the air gets colder and the sunlight hours get shorter; soon, the sun just peeks above the horizon before it sets. Eventually, over

In many areas, tundra plains are flanked by gentle hills or snowcapped mountains. This geographical variety provides habitat for an array of animals.

much of the tundra, the sun does not rise at all. The land is cloaked in darkness.

During these dark months, temperatures of 40° or 50° F below zero (−40° or −46° C) are common. The cold, dense air creates an area of high atmospheric pressure that does not easily release **moisture**. As a result, it rarely snows on the tundra during this time. From December to March, the sky over the tundra is usually clear. Stars shine brightly, and the **northern lights** shimmer like pale curtains against the deep black sky.

Although winters are long and harsh, the tundra is not without life. A few hardy animals have learned to cope with the cold. One of the most rugged year-round inhabitants of the tundra is the **musk ox**, a bulky animal with a shaggy coat and large, curved horns. Large herds of musk oxen live in some of the most desolate areas of the tundra, including the shores of the Arctic Ocean, where darkness lasts for months and the temperature never rises above zero. Even in the coldest weath-

Despite its brutally cold climate, the tundra is home to a number of plants and animals. These include tough grasses (top) and the mighty musk ox (bottom).

er, musk oxen do not seek shelter. They roam across barren hilltops, searching for frozen twigs or grass beneath the snow.

The arctic fox is another animal that can withstand the cold winters on the tundra. Like musk oxen, arctic foxes have thick, warm coats of **fur**. Their fur is gray-brown in summer but turns pure white or blue-gray in the winter. The changing colors blend in with the tundra landscape, helping the foxes avoid their enemies and surprise their prey.

Foxes eat hares, birds, and **lemmings**, which are small rodents that dig tunnels beneath the snow. When food is hard to find, the arctic fox may follow **polar bears** or wolves, which sometimes catch animals that are too big for them to eat all by themselves. After a bear or wolf has gorged itself, an arctic fox will finish off the remains.

A few types of birds spend their winters on the tundra. Ptarmigans are chickenlike birds that live in the driest areas of the tundra. They feed on tender leaves and berries in the summer but settle for frozen **vegetation** after

Arctic foxes (top) roam the dark tundra during winters in the North. The clear mid-winter sky is often illuminated by electrical activity in the atmosphere, a phenomenon known as the northern lights (bottom).

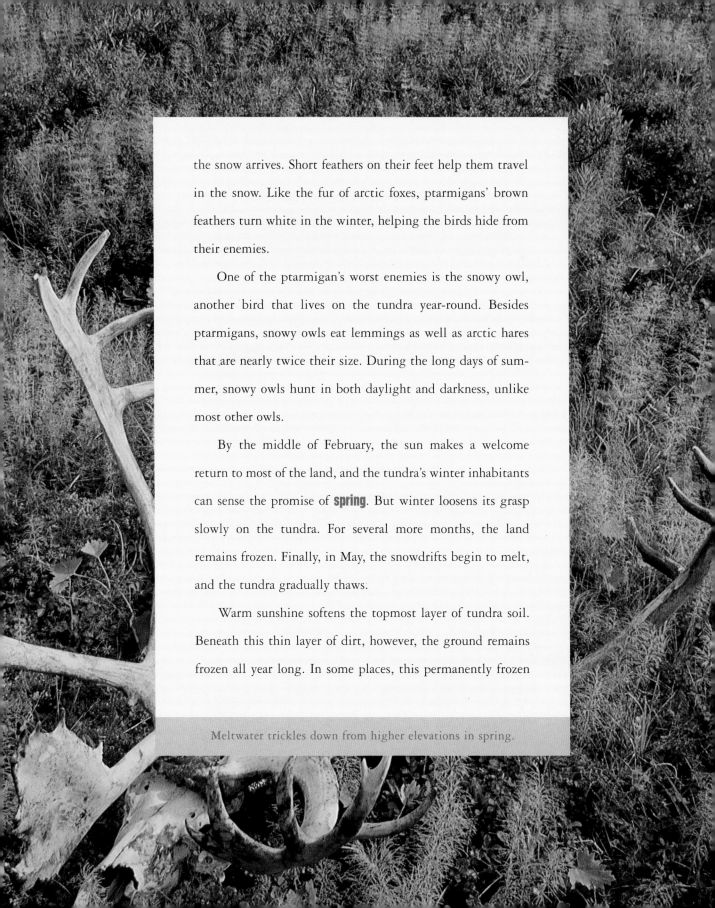

the snow arrives. Short feathers on their feet help them travel in the snow. Like the fur of arctic foxes, ptarmigans' brown feathers turn white in the winter, helping the birds hide from their enemies.

One of the ptarmigan's worst enemies is the snowy owl, another bird that lives on the tundra year-round. Besides ptarmigans, snowy owls eat lemmings as well as arctic hares that are nearly twice their size. During the long days of summer, snowy owls hunt in both daylight and darkness, unlike most other owls.

By the middle of February, the sun makes a welcome return to most of the land, and the tundra's winter inhabitants can sense the promise of **spring**. But winter loosens its grasp slowly on the tundra. For several more months, the land remains frozen. Finally, in May, the snowdrifts begin to melt, and the tundra gradually thaws.

Warm sunshine softens the topmost layer of tundra soil. Beneath this thin layer of dirt, however, the ground remains frozen all year long. In some places, this permanently frozen

Meltwater trickles down from higher elevations in spring.

ground, called **permafrost**, is nearly a mile (1.6 km) deep. Permafrost has a dramatic effect on the tundra. It prevents spring meltwater from seeping into the ground. As a result, although the tundra receives very little precipitation, each spring it becomes a wet, soggy **swamp** dotted by thousands of shallow lakes and ponds.

As summer approaches, days on the tundra gradually grow longer. By the middle of June, the sun just dips below the horizon before it rises again. Days later, the sun does not set at all. Throughout much of the tundra, the sun circles above the horizon for days or weeks. Farther north, daylight may last for months. This is why the tundra is often called "the land of the midnight sun."

Despite the long, sun-drenched days, summers on the tundra are rather cool. In July, the warmest month, the temperature usually does not rise above 50° F (10° C). Occasionally, however, there are warm days on the tundra. During a heat wave, some areas may warm up to 80° F

The tundra is full of life during the summer. Wildflowers such as larkspurs (top) rise from the shallow layer of soil, while mountain goats (bottom) and other animals on and around the tundra enjoy plentiful food.

(27° C). The mild temperatures and continuous sunshine transform the once-frozen tundra into a beautiful garden. Grasses, **sedges**, and mosses carpet the thin, water-soaked soil. The open meadows are blanketed with sprouting leaves and blooming wildflowers.

Like the tundra wildlife, tundra plants have adapted to the climate in several ways. Plants on the tundra are generally small. Summers are simply too short for trees and other plants to grow very fast. And even if the trees could grow faster, they could not secure themselves in the tundra soil. The plants' **roots** can extend only into the thin layer of sun-warmed soil; they cannot grow deeper into the permafrost.

Heather plants, such as bearberry, crowberry, and mountain cranberry, are well-adapted to the tundra **climate**. These plants have small, leathery leaves that help them conserve water during times of drought. As the plants grow, they spread outward, forming dense blankets of leaves and stems that warm the air around the plants. On a sunny day, the air next to a plant may be 25° F (14° C) warmer than the surrounding air. The

The tundra landscape, so barren during the long winter, is transformed into a beautiful garden during the summer months. Among the many short plants and wildflowers are louseworts, dwarf clover, and bearberry plants.

warmth enables the plants to grow faster.

Perhaps the most prolific tundra inhabitants are the **lichens**. Lichens are not plants, but organisms formed by algae and fungi growing together. More than 2,500 types of lichens grow on the dry areas of the tundra soil; some even grow on bare rock. Lichens vary dramatically in appearance; some look like leafless trees, while others look like miniature trumpets or patches of peeling paint.

During the long arctic winter, lichens lie **dormant** beneath the tundra snow. They start to grow as soon as the ground begins to thaw. Like tundra plants, lichens do not grow very much during the short arctic summers. A lichen as big as a baseball may be hundreds of years old. The largest lichens have been growing for thousands of years.

The summer vegetation lures many animals to the tundra. One of these summer visitors is the caribou. Caribou spend the winter in the forests below the **tree line**. As spring approaches, enormous herds of caribou leave the shelter of the forests and

Arctic ground squirrels (top) sleep through winter on the tundra, hibernating up to 10 months a year. By the time they emerge from their burrows, lichens (bottom) are growing again.

migrate north. The animals roam across the tundra, eating tiny lichens and tender grass and willows. As they forage for food, caribou must always stay alert. An unsuspecting caribou is the favorite meal of the tundra wolf.

Tundra wolves, also called **arctic wolves**, are the supreme hunters of the far North. They have excellent hearing and vision and can smell animals that are more than a mile (1.6 km) away. Large wolves can weigh more than 100 pounds (45 kg), and their strong, sleek bodies are made for running. In a sudden burst of speed, wolves can streak across the tundra at 40 miles (64 km) per hour. They also have exceptional endurance. When they are stalking prey, wolves can trot for hours without a rest.

Tundra wolves live and hunt in family groups called **packs**. People used to think that wolf packs killed caribou and other animals just for sport. It is now known that wolves kill only what they need to eat. In addition, they generally attack only young, sick, or old animals. By killing the weakest animals, wolves actually help the caribou herds. Only the

Denali National Park (opposite) in Alaska is home to many caribou and tundra wolves. These enemies actually help each other to survive; the caribou feed the wolves, and the wolves keep the caribou herds healthy.

strongest caribou survive and produce healthy offspring.

Bears also inhabit the tundra. After hibernating below the tree line in the winter, the barren ground grizzly moves north in the summer to dine on plants and tender berries. These blond-colored bears are primarily plant eaters but will also eat fish, lemmings, an occasional wolf cub, and **carrion** (dead carcasses) left by other predators.

Polar bears, on the other hand, prefer to eat meat. They live on the edge of the tundra near the Arctic Ocean, where a ready supply of **seals**, walruses, and fish can be caught. Their long bodies, powerful hindquarters, and slightly webbed toes make them great swimmers, and their thick fur insulates them from the icy Arctic water. These large predators, which can weigh 1,000 pounds (454 kg) or more, seldom wander far inland. When they do, they add berries, carrion, and other tundra animals to their diet.

Many birds, too, spend summers on the tundra. They **migrate** from warm southern lands to feed, nest, and raise their

The northernmost regions of the tundra are the realm of polar bears. These bears, which can be nearly 10 feet (3 m) tall when standing on their hind legs, are the biggest land predators in the world.

young. Millions of swans, ducks, and geese fill the ponds and marshes with ceaseless honking and quacking. Gulls, terns, and other shorebirds scurry across the tundra's rocky beaches. Falcons and eagles soar high above the tundra in search of their prey. Smaller songbirds stake out their **territories** in the low tundra hills, filling the meadows with their calls.

Most birds hatch in late June, when food on the tundra becomes plentiful. Plants begin to sprout new leaves, and the ponds and lakes are full of fish. Also about this time, an amazing number of insects pervades the tundra. The fields and meadows teem with butterflies, moths, and bumblebees. Thick clouds of **mosquitoes** and flies swarm after the caribou herds and provide food for the birds and fish.

For a time, the tundra blooms with life. But with each passing day, the sun sinks lower and lower in the sky. The meadows turn brown, and the summer visitors begin their journeys to warmer lands. By the end of August, freezing weather has returned. Soon, the Arctic tundra is gripped by frigid **darkness** once again.

Bald eagles often feed on fish plucked from lakes.

THE SUN AND EARTH

The Arctic tundra's unusual seasons are caused by the tilt of the earth as it revolves around the sun. In the summer, when the earth's northern half is tilted toward the sun, the tundra is warmed and features an abundance of wildlife. In the winter, this portion of the planet is tilted away from the sun's warming rays, bringing chilling temperatures and darkness. You can create a model to see more clearly how this happens.

You Will Need

- A flashlight
- A pencil
- A piece of paper
- A world globe

Sunlight on the Tundra

1. Have a friend hold the flashlight about one foot (30 cm) directly above the piece of paper. Use the pencil to trace around the circular spot of light.
2. Have your friend move the flashlight so that the light hits the paper at an angle. Trace around the new spot of light. You'll notice that the light pattern is oval-shaped and covers a wider area than the circular spot did.

3. Darken the room and have your friend stand close to the globe and shine the flashlight on part of the Arctic tundra—the top part of North America, for example. Position the globe so that the tundra is angled directly toward the beam of light. Notice how large the spot of light is. This is like summer on the tundra. During this season, the sun's rays strike the northern half of the earth more directly than they do the southern half. This concentrates the sun's heat on lands in the North.

4. Now position the globe so that the tundra is tilted away from the flashlight's beam but is still struck by some of the light. Again note how large the spot of light is. This is like winter on the tundra. During this season, the sun's strongest rays strike the southern half of the earth. The light that does reach the northern half is more thinly spread over a large area. As a result, the sunlight is less intense and produces little heat.

The very top of your globe may not be hit by any of the flashlight's beam at all. This is like the northernmost sections of the tundra that receive no sunlight at all for many weeks in a row during mid-winter.

ANGLING SUNLIGHT

Activity one demonstrated that sunlight strikes the earth differently depending on the time of year and the tilt of the earth. You can see the difference that the angle of sunlight makes in determining temperatures by conducting another simple experiment.

First, place a board outdoors so that its surface faces toward the sun. Then set a thermometer on the board. Read the temperature on the thermometer after a few minutes. Now tilt the board (and thermometer) so that the sun's rays strike the surface at an angle. Check the temperature again after a few minutes. You'll see that the temperature has decreased.

The sun's position in relation to the earth determines how much light and heat each part of the earth receives. This creates the different seasons. In places such as the Arctic tundra, this difference is much greater than it is in regions near the equator, which are struck by nearly the same amount of sunlight throughout the year.

LEARN MORE ABOUT THE TUNDRA

Arctic National Wildlife Refuge
U.S. Fish and Wildlife Service
101 12th Avenue, Room 236
Fairbanks, AK 99701
http://www.r7.fws.gov/nwr/arctic/arctic.html

Gates of the Arctic
(Online journal of an expedition
    across the Arctic tundra)
http://www.goals.com/thayer/gota/index.htm

Gates of the Arctic National Park
    and Preserve
National Park Service
201 First Avenue
Fairbanks, AK 99701
http://www.nps.gov/gaar/index.htm

Hudson Bay Project
(Online study of the coastal Arctic tundra)
http://research.amnh.org/~rfr/hbp/main.html

North American Lichen Project
(Online resource for information
    about lichens)
http://www.lichen.com/

Polar Bears
SeaWorld Education Department
http://www.seaworld.org/polar_bears/
    pbindex.html

Tundra Animal Printouts
(Online resource for information
    about tundra animals)
http://www.zoomwhales.com/biomes/
    tundra/tundra.shtml

INDEX

Bearberry plants are one of many unique tundra inhabitants.